Nutritional Importance in Human Epigenome Mechanisms

Muhammad Mazhar Fareed

GRIN

Bibliographic information published by the German National Library:

The German National Library lists this publication in the National Bibliography; detailed bibliographic data are available on the Internet at http://dnb.dnb.de.

ISBN: 9783346571649
This book is also available as an ebook.

Print and binding: Books on Demand GmbH, Norderstedt, Germany
Printed on acid-free paper from responsible sources.

GRIN web shop: https://www.grin.com/document/1164063

Nutritional Importance in Human epigenome mechanisms

Author:

[1]Muhammad Mazhar Fareed

[1]Faculty of Life Sciences, Department of Bioinformatics and Biotechnology, Government College University, Faisalabad, Pakistan

Date: 22/09/2021 to 01/12/2021

Abstract: Environmental epigenetics depicts how natural elements influence cell epigenetics what's more, consequently, human wellbeing. Epigenetic marks modify the spatial conformity of chromatin to direct quality articulation. Ecological variables with epigenetic impacts incorporate practices, nourishment, furthermore, synthetic compounds and modern contaminations. Epigenetic systems are additionally embroiled during improvement in utero and at the cell level, so natural openings might hurt the baby by weakening the epigenome of the creating organic entity to change illness hazard further down the road. Paradoxically, bioactive food parts might trigger defensive epigenetic adjustments all through life, with right on time life sustenance being especially significant. Past their hereditary qualities, the general wellbeing status of an individual might be viewed as a coordination of numerous natural signs beginning at incubation furthermore, acting through epigenetic changes. This audit investigates how the climate influences the epigenome in wellbeing and infection, with a specific spotlight on the disease. Understanding the atomic impacts of conduct, supplements, and poisons may be applicable for creating protection techniques also, customized health programs. Moreover, by re-establishing cell separation, epigenetic drugs could address a likely procedure for the therapy of numerous sicknesses including cancer/malignant growth.

Key words: Environmental epigenetics, cell level, nutrients, cancer

1.Introduction

Epigenetics alludes to the control of quality articulation of genes through instruments not straightforwardly identified with the DNA coding grouping. Therefore, all cells in a creature have totally different aggregates notwithstanding having a similar genome. Epigenetics adjusts and directs quality articulation through different epigenomic "marks", the term given to synthetic mixtures added to DNA or histone proteins and perceived by catalysts that either set down or eliminate the particular imprint (Grunstein, 1997; Guerrero-Bosagna & Skinner, 2012). These imprints change the spatial conformity of chromatin: either compacting it, subsequently forestalling the limiting of record elements to the DNA, or opening it, permitting record factor restricting and generally upregulating cell processes. DNA methylation is the expansion of methyl gatherings to the 5-carbons of cytosine deposits in CpG islands to bring about 5-methylcytosines-cooperates with histone changes to control quality articulation (Struhl, 1998). DNA methylation will in a general demonstration at advertisers to prompt quality hushing, while histone acetylation normally loosens up chromatin. DNA methylation is catalyzed by chemicals in the DNA methyltransferase (DNMT) family, which enrols the utilitarian buildings concerns containing DNA methylation areas, prompting transcriptional hindrance or keeping an abusive chromatin state (Kouzarides, 1999).

Histone acetylation is related to transcriptional action and an open chromatin state. Acetylation of histone tails is constrained by two compound families: histone acetyltransferases (HATs), which move an acetyl gathering, and histone deacetylases (HDACs), which eliminate acetyl gatherings. Other epigenetic marks are additionally depicted and incorporate histone post-translational alterations such as methylation, ubiquitination, sumoylation, phosphorylation, biotinylation, and ADP-ribosylation, which either advance or stifle quality articulation. The example of these imprints on histone tails is frequently alluded to as the histone code, which directs the limiting of effector proteins that thus results in explicit cell processes. Non-coding RNAs (ncRNAs) are one more kind of explicit epigenetic mark that intervene differently intracellular cycles. An ncRNA is a useful RNA atom interpreted from DNA yet not converted into protein. The best-described ncRNAs are microRNAs (miRNAs), which are short, single-abandoned, 19–24 nucleotide ncRNAs. miRNAs direct quality hushing at the transcriptional and additionally the translational degree of protein-coding qualities (Huang, Jiang, & Zhang, 2014). Long non-coding RNAs (lncRNAs) are another subset of RNAs more than 200 nucleotides long that capacity as chromatin remodelers, transcriptional controllers, and post-transcriptional controllers. Numerous lncRNAs complex with chromatin-adjusting proteins to enlist their reactant action to explicit genomic locales, in this manner adjusting chromatin states and impacting quality articulation. With almost no protein-coding potential, lncRNAs all things considered take part in different intracellular cycles, and ongoing examinations have recognized that specific lncRNAs are explicitly connected with specific malignant growths. With the improvement of high-goal sequencing and high-throughput advancements, a huge number of naturally useful ncRNAs have now been recognized (Lillycrop, Hoile, Grenfell, & Burdge, 2014). Likewise, there is crosstalk between DNA methylation and histone adjustments; for instance, histone methyltransferases, histone demethylases and embellishment proteins communicate and arrange the chromatin state and DNA methylation and methylation status of histones are firmly related. Besides, different ncRNAs are likewise firmly connected with other epigenetic marks, which structure broad crosstalk all through the cell, or the "epigenetic network" (Choi & Friso, 2010; Hussey, Lindley, & Mastana, 2017). There is presently a lot of proof that the epigenetic control of the genome is definitely more convoluted than first suspected and includes different epigenetic systems and their associations. All the more as of late, histone variations, for the most part of sanctioned histones H2A, H2B, and H3, with explicit properties have been distinguished in people and other higher eukaryotes. The vast majority of them are H2A variations, among which macroH2A1 produces elective graft isoforms, i.e., macroH2A1.1 and macroH2A1.2. MacroH2A1 isoforms give off an impression of being basic controllers of chromatin structure and chromatin elements during cell senescence, recovery, and fasting. The capacity of these histone variation proteins and their atomic components in wellbeing and during an organic entity's life expectancy is investigated. The revelation of histone variations and their variety has added further intricacy to setting subordinate organic frameworks and their guideline, for example, during wellbeing, maturing, and in obsessive conditions including malignant growth. This is a space of continuous examination since chromatin elements all through life are probably going to change and the epigenome experiences a moderate misfortune in setup during maturing. The subsequent strange chromatin state during maturing is described by various consolidated histone variations, nucleosome redesigning, modified histone alteration designs, and modified DNA methylation designs, bringing about the enlistment of various chromatin modifiers, unusual quality articulation designs, and genomic precariousness (Lo Re & Vinciguerra, 2017). Among the various variations of histone H2A, macroH2A has been embroiled in the maturing (Pal & Tyler, 2016); its level expansions in the age-subordinate way during replicative senescence in refined human fibroblast cells and furthermore in a few tissues of matured mice and primates.

2.Methodologies

The research based data were obtained from the different tool-based research sites on environmental, epigenetics changes, nutritional values and factors-health and medical based information as; NIH (National Institutes of Health (NIH) | Turning Discovery Into Health), PubMed (PubMed (nih.gov)), Medical Research Council (Home - Medical Research Council (ukri.org)), Nutritional-Research (HOME (nutritionalresearch.org), NIH-Health_Enviornment (Epigenetics (nih.gov)), after that all results are compiled and analysed.

3.Environmental impact of epigenetic

These dynamic or oppressive imprints are additionally reliant upon the way of life and natural elements. "Ecological epigenetics" alludes to how natural openings influence epigenetic changes. Educational encounters, propensities, and our current circumstances shape what and who we are by temperance of their effect on our epigenome and wellbeing; for example, albeit indistinguishable twins share a similar genome and are hastily phenotypically comparative, they are one-of-a-kind people with determinable contrasts (Pal & Tyler, 2016). These distinctions result from unmistakable quality articulations affected by epigenetic factors. Conduct, nourishment and openness to poisons and toxins are among the way of life factors known to be related with epigenetic changes. For instance, nourishment is a vital ecological openness from incubation to death that impacts our wellbeing by affecting epigenetic peculiarities. In another model, later epidemiological information recommends that the expanded frequency of malignant growth seen in the created world since the 1960s may mostly be because of openness to endocrine-disturbing synthetics (EDCs), to which people and natural life are uncovered day by day from numerous sources (Pal & Tyler, 2016).

4.Cancer perspectives

Cancer is the second most normal reason for death in many nations and will remain so as older individuals are generally helpless to malignant growth and the populace is quickly maturing, essentially in the west. While age-normalized malignant growth death rates are projected to diminish in the European Union (EU) and the United States (US) because of advances in screening, anticipation, and treatment, the rate has expanded in Europe and the US for testicular and prostate malignant growths in the course of the last 50 years (Malvezzi et al., 2015; Reamon-Buettner, Mutschler, & Borlak, 2008). In the United Kingdom (UK), the joined occurrence of all malignant growths has expanded for all age bunches since the mid-1990s, with the best increment found in youngsters, matured 0–24 years. Malignant growth is a hereditary sickness portrayed by acquired or inconsistent changes in qualities that keep up with tissue homeostasis, control the cell cycle or manage apoptosis. Malignant growth is additionally an epigenetic infection portrayed by transformations in chromatin-renovating proteins and epigenome modifications coming about from distorted connection or evacuation of DNA or histone protein marks (DeSantis et al., 2014; Strohsnitter et al., 2001). Gathering proof recommends that numerous grown-up infections, including tumors, have epigenetic beginnings.

5.Nutritional factors effecting in Health with epigenetics

Nutrition is one of the most considered and better comprehended ecological epigenetic factors, also affiliations have been seen between unfriendly pre-birth nourishing conditions, post-pregnancy wellbeing, what's a more, expanded danger of sickness (Argente, Mehls, & Barrios, 2010). For example, at its limit, the Dutch Famine Birth Cohort coming about from the Dutch Famine of 1944–1945 has been utilized to concentrate on the impacts of starvation during pregnancy also, ensuring wellbeing and formative results including, yet not restricted to, expanded danger of type II diabetes mellitus, cardiovascular infection, metabolic issues, and diminished intellectual capacity in later life (Barker & Clark, 1997). The primary long periods of pregnancy appear to have the best impact on infection hazards; youngsters considered during the Dutch Famine would in general have more modest than-regular posterity, proposing the impacts as might continue, affect our youngsters and even past. It appears to be possible that the embryo epigenetically adjusts because of a restricted stockpile of supplements. In people, industrious epigenetic contrasts related with pre-birth openness to starvation have been attributed to a lower level of methylation of a quality ensnared in insulin digestion than their unexposed kin. The proof for transgenerational impacts of poor maternal eating regimen on human populaces with regard to metabolic results was inspected. There is proof from verifiable records that the grandkids of ladies presented to starvation and other dietary modifications during pregnancy are bound to encounter unexpected problems than their control partners (Knopik, Maccani, Francazio, & McGeary, 2012). The potential sub-atomic components of transgenerational legacy recommend the methylation of gametes by means of the fatherly and maternal ancestry. For sure, further transmission through the fatherly line is exceptionally liable to happen through epigenetic regulation of the spermatozoan core. Two models from verifiable accomplices delineate this transgenerational transmission through fatherly heredity.

One showed that female grandkids (F2) from the fatherly grandma (F0) who experienced helpless food accessibility during her own development was at higher danger of cardiovascular mortality. The second later model showed that grown-up amazing posterity whose fathers were presented to starvation in utero had higher BMIs then a controlled populace. The proof that both maternal and fatherly eating regimens impact metabolic aggregates in posterity in vertebrates through epigenetic data transmission is surveyed (Heijmans et al., 2008).

Over sub-atomic systems as for the fetal beginnings of the grown-up illness have been proposed counting mitochondrial brokenness and oxidative pressure as among the soonest occasions depicted in posterity presented to supplement limitation. Sustenance in early life incites long haul changes in DNA methylation that effect on person wellbeing and age-related infections all through life. Supplements can either act straight by hindering epigenetic catalysts like DNMT, HDAC, or HAT or by changing the accessibility of substrate essential for those enzymatic responses. This thus adjusts the outflow of basic qualities and effects our general wellbeing and life span. Various investigations have revealed the epigenetic impacts of diet on aggregate and helplessness to illnesses all through life. Folate digestion is connected to phenotypic changes through DNA methylation, as folate, a water-dissolvable B nutrient is a wellspring of one-carbon for the union of AdoMet, which is essential for DNA methylation. Other methyl giver supplements, for example, choline can likewise change the DNA methylation status and consequently sway quality articulation. Maternal methyl contributor supplement accessibility in early pregnancy is fundamental for the legitimate fetal turn of events, with results for wellbeing and infection defencelessness or malignant growth in the kids all through life. In one creature study, a maternal eating routine confined in methyl supplements during bias impacted DNA methylation designs in posterity and was the reason for changed aggregates. On the other hand, dietary limitation yet without serious healthful hardship has been displayed in a few models to broaden life expectancy (Aiken, Tarry-Adkins, & Ozanne, 2016; Rando & Simmons, 2015). Calorie limitation has a mitigating impact through the inhibitory impacts of basic qualities not restricted to NF-κB. From the epigenetic viewpoint, there is clear crosstalk between DNA methylation and histone changes, recommending that the chromatin construction may likewise decide DNA methylation. In this setting of epigenetic associations, sirtuin 1, a NAD+ subordinate HDAC whose substrate explicitness incorporates histone proteins, has been proposed to be enacted by some dietary parts (for instance resveratrol, a kind of regular phenol present in grape skins). Sirtuin 1 intercedes a portion of the impacts of dietary limitation that deferral or opposite a portion of the physiological changes related with maturing through consequences for DNA methylation. Eats less carbs high in these methyl-giving supplements can quickly affect quality articulation, particularly during early improvement when the epigenome is first settled and can have dependable impacts in grown-up life. Studies in creatures have announced that consumes less calories poor in methyl-giving folate or choline previously or soon after birth cause extremely durable hypomethylation of parts of the genome. In grown-ups, a methyl-insufficient eating regimen diminishes DNA methylation, however, the progressions are reversible when methyl is added once more into the eating regimen. Further, contingent upon the dietary enhancements got by a pregnant mouse, her posterity might have an alternate aggregate because of differential quality methylation. In a model of epigenome-adjusting synthetic compounds, bisphenol A (BPA) is generally used to fabricate variously plastic items including compartments. The puppies of grown-up mice took care of BPA were bound to have an unfortunate aggregate (yellow/fat, inclined to malignant growth and diabetes) contrasted with those brought into the world from moms who took care of BPA with enhanced methyl-rich supplements like folic corrosive and nutrient B12 (brown, slim and solid) (Kim, Friso, & Choi, 2009; Rando & Simmons, 2015). For this situation, maternal supplement supplementation balanced the adverse consequences of substance openness, highlighting the significance of a decent eating regimen wealthy in leafy foods and other great food sources. Methyl-giving supplements go about as co-substrates for methyl bunch moves; the pool of accessible methyl contributors is a significant controller of both DNA and histone methylation limit and their creation are additionally subject to BPA's epigenetic impacts (Davis & Ross, 2007). In this mouse model, changed DNA hypomethylation could be eased by folic corrosive as a dietary methyl benefactor, and also summarizes some dietary components considered to have protective effects against cancer through different epigenetic modifications in (Table 01).

Table No 1: A summary of dietary components for cancer mechanisms

No.	Dietary Nutrients	Role of epigenetics
1	Methionine	SAM synthesis
2	Folic acid	Methionine synthesis
3	Vitamin B-12,06	Methionine synthesis
4	Resveratrol	Removes acetyl groups from histones
5	Genistein	Methylation, cancer

		prevention, mechanism
6	Sulforaphane	Histone acetylation
7	Butyrate	Break down the toxic by-products

One more striking illustration of the impacts of early eating regimens on epigenetics with results on the aggregate can be found in bumblebees. The sterile working drone separates from the rich sovereign relying upon the larval eating routine through epigenetic changes in DNA methylation designs. Hatchlings assigned to become sovereigns are taken care of solely with imperial jam, which contains epigenetically dynamic fixings that quiet a key quality which itself hushes a gathering of sovereign qualities (Davis & Ross, 2007; Ford, Ions, Alatawi, & Wakeling, 2011). Moreover, DNA methylation changes happen during maturing, and it has become apparent that early life nourishment can tweak DNA methylation and impact life span, specifically by actuating long haul changes in DNA methylation and different imprints that influence defencelessness to a scope of maturing related illnesses.

6.Nutritional effect on epigenetic

Folic acid and vitamin B12 are two examples of epigenetically dynamic fixings that play significant jobs in DNA digestion and the support of DNA methylation designs through the substance response of a methyl item. In one in vivo study, dietary folate admission was decidedly related with p16 growth silencer quality articulation, a basic disease-related quality with regular hushing DNA methylation of its advertiser (Alavian-Ghavanini & Rüegg, 2018; Ford et al., 2011). Changed p16 quality articulation was seen in matured mouse colons, reliable with the known decline in DNMT articulation with maturing. Low folate admission has been related to hypomethylation and an expanded danger of colorectal furthermore, pancreatic diseases. There is a developing assortment of epidemiological proof that folate tweaks anticarcinogenic properties through epigenetic changes, as folate lack decreases the potential for DNA methylation, and strange DNA methylation is related with many sorts of disease. Eats less carbs wealthy in products of the soil containing regular enemies of oxidants can secure against disease (Bishop & Ferguson, 2015; Larsson, Giovannucci, & Wolk, 2006).

The potential epigenetic impacts of a few dietary parts notwithstanding folate, are generally determined from vegetables, have been analyzed in various investigations, which have shown decreases in DNA hypermethylation of basic qualities bringing about growth concealment. For instance, green tea contains polyphenols, which are normal mixtures broadly dispersed in plant food varieties and with numerous organic exercises including hindrance of DNA methylation. An assortment of dietary elements are potential HDAC and HAT modulators. A few, for example, sulforaphane, an isothiocyanate found in broccoli sprouts, or diallyl disulfide, an organosulfur compound in garlic, have been displayed to go about as HDAC inhibitors, a class of epigenetic helpful portrayed further beneath. Such epigenetic drugs have been utilized to treat malignant growths in clinical preliminaries because of their method of activity in re-establishing malignant growth cell separation and delivering cancers more delicate to customary treatments (Dashwood & Ho, 2007; Druesne et al., 2004). A few in vitro concentrates on utilizing these mixtures have shown hostile to cancer-causing impacts related with HDAC hindrance and histone acetylation. As indicated above, miRNAs can direct DNA methylation and histone changes, yet advertiser methylation or histone acetylation can likewise regulate miRNA articulation as a feature of a mind-boggling network with feed-forward and input circles (Fabiani, Minelli, Bertarelli, & Bacci, 2016). Dysregulated miRNA articulation is related with the improvement or movement of human malignant growths through changes in cell expansion and apoptosis, yet, methyl-and folate-inadequate eating regimens can likewise bring about abnormal miRNA articulation to apply comparably, favourable to disease impacts. Certain dietary parts might secure against disease through miRNA guideline, such as; curcumin and retinoic corrosive the previous presence in certain plants, what's more, normally utilized as a dietary enhancement and food seasoning, and the last option is present in any nutrient A-rich food (Fung et al., 2003; Jensen et al., 2018). And summarizes how diet affects the epigenome to modify individual and transgenerational phenotypes in Figure no 1.

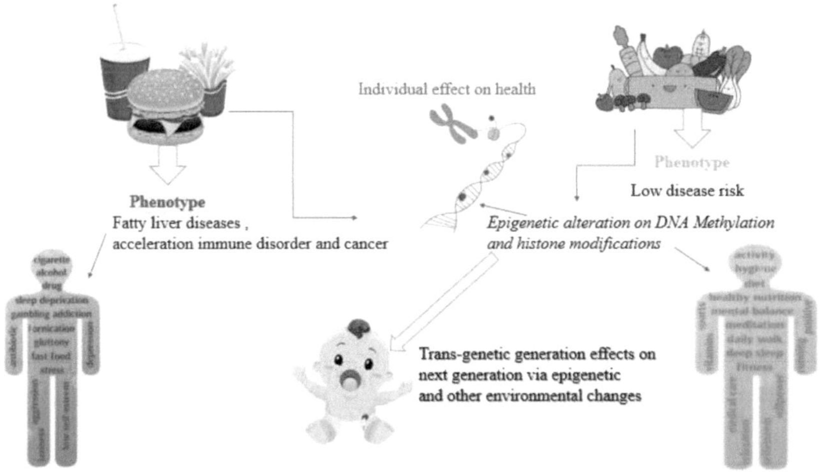

Figure No 01: Diet affects the epigenome to modify individual and transgenerational phenotypes

7.Environmental changes in epigenetics along chemically

Environmental pollutants are among the natural or environmental variables with epigenetic impacts. Natural openings also called environmental exposures regularly considered epigenetic poisons incorporate metals, like arsenic, particulate matter from air contamination, natural mixtures, pesticides, and other endocrine-disturbing synthetic compounds address a gathering of ecological synthetic substances that meddle with chemical flagging pathways (Alavian-Ghavanini & Rüegg, 2018). Their activities on the epigenome are depicted as a model. Peoples are routinely presented to a wide scope of Endocrine-disrupting chemicals (EDCs). There is expanding proof to propose that EDC openness during early life, specifically during fetal turn of events, adds to an assortment of sicknesses, including tumors, which show further down the road or even in the following age. This recommends the epigenetic legacy of EDC-related occasions. To be sure, it is grounded that epigenetic instruments assume a focal part in these dependable impacts of EDCs. In addition, an enormous number of studies have shown that EDCs prompt epigenetic changes. Notwithstanding, the basic instruments of activity are less surely known (Knopik et al., 2012). The impacts of EDCs on epigenetics were as of late assessed. Momentarily, the impacts of EDCs can be worldwide, that is, on epigenetic catalyst articulation, i.e., DNMTs, HATs, and HDACs. For this situation, for instance, EDCs influence

DNMT by means of atomic chemical receptors like the estrogen and androgen receptors or through miRNA articulation, the last option being accompanying with other quality articulation changes. The second method of activity of EDCs is quality explicit, directing locus-explicit epigenetic designs. For instance, EDCs may explicitly influence one quality by means of atomic receptors, which have been answered to control locus-explicit chromatin states by enrolling histone modifiers and evolving DNA methylation designs by associating with DNMTs and enrolling them to explicit genomic locales (Knopik et al., 2012).

Moreover, EDCs are probably going to manage other ncRNA articulations by means of obstruction with atomic receptors, which play a putative part in the guideline of ncRNA articulation. While smoking and liquor utilization predominantly cause infection through mutagenic occasions, both can likewise apply cell impacts through epigenetics. Kids presented to pre-birth maternal smoking have epigenetic changes in buccal mucosal cells (Doherty, Grabowski, Hoffman, Ng, & Zelikoff, 2009). Maternal cigarette smoking during pregnancy has been researched regarding outcomes on wellbeing, neurodevelopment, conduct, and intellectual work in new-born children. Maternal tobacco smoking during pregnancy is related to respiratory illnesses related with safe framework dysregulation, for example, asthma and sensitivities just as malignant growth later in the existence of posterity. In another model, liquor use is a danger factor for some, diseases including oral, pharyngeal, laryngeal, oesophageal, liver, colorectal, and bosom disease, and the danger of malignant growth is portion subordinate. Liquor has injurious/unsafe consequences for development, digestion, and neuronal advancement through setting down epigenetic marks at early-stage openness, as proven by in vitro and creature studies. These unsettling influences in the epigenetic scene might add to deserts in a specific memory and learning processes just as mutations and unusual fetal turn of events. Various examinations have detailed that both hereditary and ecological danger factors additionally assume a part in the improvement of liquor abuse (Ducci & Goldman, 2008; Farris, Wolen, & Miles, 2010).

8.Conclusion and Future perspectives

The wellbeing of an individual relies upon the connection of numerous ecological elements with its hereditary qualities. Way of life factors is probably going to influence human wellbeing and posterity by means of epigenetic components. Varieties in quality articulation are impacted by epigenetics, which relies upon our background also, propensities include sustenance, conduct, and ecological poison openings. Constant ecological openings are believed to be incompletely answerable for the expanded paces, all things considered, not just for those straightforwardly uncovered, yet additionally for the hatchling. A few times of advancement are especially helpless to the hurtful impacts of these poisons. Also, the mix of poison, portion, and basic openness a window should be thought about yet might be hard to anticipate. Following poison openness, epigenomic changes in the posterity might influence a few organs and make the individual touchier to cancer-causing agents during youth or adulthood and increment illness defencelessness sometime down the road, including disease, diabetes, and chemical imbalance. Epigenetic marks are impacted by the climate and these progressions might be restored in resulting age, or transgenerational epigenetic legacy. Some of our qualities, practices, sicknesses, and great and awful encounters leave epigenetic labels that might be acquired; nonetheless, they may likewise be eliminated by way of life changes. There is expanding proof that distinctive epigenetic components (DNA methylation, histone adjustments, ncRNAs) are interconnected and structure an "epigenetic network”. The way of life factors and ecological openings leave epigenetic marks on our DNA that sways quality articulation; some have defensive impacts while others are unsafe. Foods grew from the ground vegetable-rich weight control plans might effectively affect DNA to epigenetic drugs. A more profound agreement of epigenetic impacts and the flagging pathways enacted by bioactive food parts would help in evaluating the job and likely advantage of supplements on our wellbeing and decreasing disease powerlessness. Nourishing epigenetics could be joined with drugs for synergistic impacts for therapy or prophylaxis or be adjusted for a pregnant lady to diminish the weight of persistent illness in posterity through an "epigenetically sound" gestational eating routine. In both created and creating. nations, advancing maternal eating regimen is a difficult general medical issue. Future work in the field of nourishment and epigenetics can possibly give critical advantage to general wellbeing, also, customized sustenance could turn into a piece of a patient's clinical program. The equivalent applies to ecological synthetics. Because of their unsafe impacts on people, youngsters, and perhaps across ages, the effect of synthetic compounds with epigenetic impacts has turned into a major worldwide general wellbeing worry that should be tended to. A few designated spots or key particles in the cell at the interface between the cell cycle and separation processes are probably going to be targets of synthetics that favor malignant growth advancement. Understanding the epigenetic impacts of EDCs would help in evaluating their activity and planning such stamps could be utilized for hazard the executives dependent on pathway-based harmfulness testing.

References

Aiken, C. E., Tarry-Adkins, J. L., & Ozanne, S. E. (2016). Transgenerational effects of maternal diet on metabolic and reproductive ageing. *Mammalian Genome, 27*(7), 430-439.

Alavian-Ghavanini, A., & Rüegg, J. (2018). Understanding epigenetic effects of endocrine disrupting chemicals: from mechanisms to novel test methods. *Basic & clinical pharmacology & toxicology, 122*(1), 38-45.

Argente, J., Mehls, O., & Barrios, V. (2010). Growth and body composition in very young SGA children. *Pediatric Nephrology, 25*(4), 679-685.

Barker, D. J., & Clark, P. M. (1997). Fetal undernutrition and disease in later life. *Reviews of reproduction, 2*, 105-112.

Bishop, K. S., & Ferguson, L. R. (2015). The interaction between epigenetics, nutrition and the development of cancer. *Nutrients, 7*(2), 922-947.

Choi, S.-W., & Friso, S. (2010). Epigenetics: a new bridge between nutrition and health. *Advances in nutrition, 1*(1), 8-16.

Dashwood, R. H., & Ho, E. (2007). *Dietary histone deacetylase inhibitors: from cells to mice to man.* Paper presented at the Seminars in cancer biology.

Davis, C. D., & Ross, S. A. (2007). Dietary components impact histone modifications and cancer risk. *Nutrition reviews, 65*(2), 88-94.

DeSantis, C. E., Lin, C. C., Mariotto, A. B., Siegel, R. L., Stein, K. D., Kramer, J. L., . . . Jemal, A. (2014). Cancer treatment and survivorship statistics, 2014. *CA: a cancer journal for clinicians, 64*(4), 252-271.

Doherty, S., Grabowski, J., Hoffman, C., Ng, S., & Zelikoff, J. (2009). Early life insult from cigarette smoke may be predictive of chronic diseases later in life. *Biomarkers, 14*(sup1), 97-101.

Druesne, N., Pagniez, A., Mayeur, C., Thomas, M., Cherbuy, C., Duée, P.-H., . . . Chaumontet, C. (2004). Diallyl disulfide (DADS) increases histone acetylation and p21 waf1/cip1 expression in human colon tumor cell lines. *Carcinogenesis, 25*(7), 1227-1236.

Ducci, F., & Goldman, D. (2008). Genetic approaches to addiction: genes and alcohol. *Addiction, 103*(9), 1414-1428.

Fabiani, R., Minelli, L., Bertarelli, G., & Bacci, S. (2016). A western dietary pattern increases prostate cancer risk: a systematic review and meta-analysis. *Nutrients, 8*(10), 626.

Farris, S. P., Wolen, A. R., & Miles, M. F. (2010). Using expression genetics to study the neurobiology of ethanol and alcoholism. *International review of neurobiology, 91*, 95-128.

Ford, D., Ions, L. J., Alatawi, F., & Wakeling, L. A. (2011). The potential role of epigenetic responses to diet in ageing. *Proceedings of the Nutrition Society, 70*(3), 374-384.

Fung, T., Hu, F. B., Fuchs, C., Giovannucci, E., Hunter, D. J., Stampfer, M. J., . . . Willett, W. C. (2003). Major dietary patterns and the risk of colorectal cancer in women. *Archives of internal medicine, 163*(3), 309-314.

Grunstein, M. (1997). Histone acetylation in chromatin structure and transcription. *nature, 389*(6649), 349-352.

Guerrero-Bosagna, C., & Skinner, M. K. (2012). Environmentally induced epigenetic transgenerational inheritance of phenotype and disease. *Molecular and cellular endocrinology, 354*(1-2), 3-8.

Heijmans, B. T., Tobi, E. W., Stein, A. D., Putter, H., Blauw, G. J., Susser, E. S., . . . Lumey, L. (2008). Persistent epigenetic differences associated with prenatal exposure to famine in humans. *Proceedings of the National Academy of Sciences, 105*(44), 17046-17049.

Huang, B., Jiang, C., & Zhang, R. (2014). Epigenetics: the language of the cell? *Epigenomics, 6*(1), 73-88.

Hussey, B., Lindley, M. R., & Mastana, S. (2017). Epigenetics and epigenomics: the future of nutritional interventions? : Future Science.

Jensen, V. S., Hvid, H., Damgaard, J., Nygaard, H., Ingvorsen, C., Wulff, E. M., . . . Fledelius, C. (2018). Dietary fat stimulates development of NAFLD more potently than dietary fructose in Sprague–Dawley rats. *Diabetology & metabolic syndrome, 10*(1), 1-13.

Kim, K.-c., Friso, S., & Choi, S.-W. (2009). DNA methylation, an epigenetic mechanism connecting folate to healthy embryonic development and aging. *The Journal of nutritional biochemistry, 20*(12), 917-926.

Knopik, V. S., Maccani, M. A., Francazio, S., & McGeary, J. E. (2012). The epigenetics of maternal cigarette smoking during pregnancy and effects on child development. *Development and psychopathology, 24*(4), 1377-1390.

Kouzarides, T. (1999). Histone acetylases and deacetylases in cell proliferation. *Current opinion in genetics & development, 9*(1), 40-48.

Larsson, S. C., Giovannucci, E., & Wolk, A. (2006). Folate intake, MTHFR polymorphisms, and risk of esophageal, gastric, and pancreatic cancer: a meta-analysis. *Gastroenterology, 131*(4), 1271-1283.

Lillycrop, K. A., Hoile, S. P., Grenfell, L., & Burdge, G. C. (2014). DNA methylation, ageing and the influence of early life nutrition. *Proceedings of the Nutrition Society, 73*(3), 413-421.

Lo Re, O., & Vinciguerra, M. (2017). Histone MacroH2A1: a chromatin point of intersection between fasting, senescence and cellular regeneration. *Genes, 8*(12), 367.

Malvezzi, M., Bertuccio, P., Rosso, T., Rota, M., Levi, F., La Vecchia, C., & Negri, E. (2015). European cancer mortality predictions for the year 2015: does lung cancer have the highest death rate in EU women? *Annals of Oncology, 26*(4), 779-786.

Pal, S., & Tyler, J. (2016). Epigenetics and aging. Sci Adv 2: e1600584.

Rando, O. J., & Simmons, R. A. (2015). I'm eating for two: parental dietary effects on offspring metabolism. *Cell, 161*(1), 93-105.

Reamon-Buettner, S. M., Mutschler, V., & Borlak, J. (2008). The next innovation cycle in toxicogenomics: environmental epigenetics. *Mutation Research/Reviews in Mutation Research, 659*(1-2), 158-165.

Strohsnitter, W. C., Noller, K. L., Hoover, R. N., Robboy, S. J., Palmer, J. R., Titus-Ernstoff, L., . . . Hatch, E. E. (2001). Cancer risk in men exposed in utero to diethylstilbestrol. *Journal of the National Cancer Institute, 93*(7), 545-551.

Struhl, K. (1998). Histone acetylation and transcriptional regulatory mechanisms. *Genes & development, 12*(5), 599-606.